The Lockdown Pallet Hive

Jonathan Powell

The Lockdown Pallet Hive
ISBN: 978-1-914934-14-8
Text, Graphics and Photos: Jonathan Powell
Published by Northern Bee Books 2021
Northern Bee Books, Scout Bottom Farm,
Mythomroyd, Hebden Bridge, HX7 5JS (UK)
www.northernbeebooks.co.uk
Tel: +44 (0) 1422 882751

Designed by: Lynnette Busby
www.whatever.graphics

The Lockdown Pallet Hive

Jonathan Powell

Contents

Preface.

The Lockdown Pallet Hive is written by Jonathan Powell, a Trustee of the Natural Beekeeping Trust (www.naturalbeekeepingtrust.org) and a beekeeper since he was a small boy.

Jonathan has a keen interest in Traditional Tree Beekeeping and is the author of "The Field Guide to Tree Beekeeping". He is also a host of the "Arboreal Apiculture Salon" podcast, and regular presenter on topics regarding Apiology.

He oversees honey bee rewilding projects at organic farms in the UK and Spain.

Summer 2020

The World has Changed

We often get asked at the Natural Beekeeping Trust - "Can you make a hive from pallets?" It's a good question because hives can often cost more than £400 and yet we often have trouble disposing of these wood pallets that clutter our back yards and gardens. Unable to resist a challenge, I set about making the ultimate pallet hive incorporating all my knowledge about tree beekeeping, wild free-living bees and their preferences. This is a hive just for bees, and not for the beekeeper interested in "a little honey".

I call this a Lockdown Hive because the coronavirus lockdown forced me to look for hive materials around the garage, otherwise I would have nowhere to welcome this year's swarms after disposing of my box hives. The only materials I had at hand were a couple of pallets."

The Pallet

Figure 1 Standard Euro Pallet

The first thing to understand about building a pallet hive is that pallet wood comes in all sorts of different sizes. So, this article is very much a guide and not a plan. Alan Wood has created an extremely useful excel calculator for this project that can be downloaded from this URL: https://www.naturalbeekeepingtrust.org/post/the-lockdown-pallet-hive. With the calculator you can vary the hive volume, number of sides, cavity size and wood dimensions to suit your pallets and hive design goals.

Never use a pallet with the marking MB on it as these have been treated with Methyl Bromide (outlawed in UK since 2010). Use pallets which have been heat treated and have HT, KD or DH markings.

For more information visit Universal Pallets:
https://www.universalpallets.com/2018/01/ultimate-guide-pallet-markings/
This is a wonderful website full of pallet nerdery.

To make a hive, the pallets must be good quality, untreated, and ideally use 10-20cm wide planks that are 2cm thick. Thinner, lighter wood could be used but this would need to be compensated for with more insulation. Lighter hives also don't have much thermal inertia- a property that smooths out the night and day temperature fluctuations in the hive, thus reducing hive stress.

Tools & Deconstruction

You will need two pallets, a table saw, general woodworking tools and two days to build the hive in this project. I have found that any hive worthy of bees (Sun Hive, Lazutin Hive, Zeidler Hive, log hive) takes two days. It's like a golden rule of hive making for me.

Figure 2 A disassembled pallet with annular nails removed - my least favourite part of the build

Above is the first of two pallets used in this project. Except for the nails, we will use everything from the pallets; even the sawdust will be used for the hive insulation.

Beginning Construction

First glue the pallet separator blocks together using a waterproof wood glue. Be careful not to over glue as many types of wood glue contain biocides and we don't want that in contact with the bees.

Figure 3 Creating the roof and floor

Above I have glued the blocks together and then sawn them in half to make a top and bottom 'plug' for the hive. (I had to add an extra shim of wood to get the width I wanted). They are both about 5cm thick. I have chosen an octagon because that is the minimum shape to avoid cold corners, but nonagons, decagons, undecagons ... would all approximate better to a circle, it's just more work. If you have narrow planks, you may need more faces to ensure you have the minimum required internal diameter. In addition, extra sides have the bonus of creating more joint cracks for bees to seal with propolis. Bees use the propolis together with the hive warmth and humidity to create a sterile medicinal atmosphere.
Without good insulation the hive atmosphere cannot be easily maintained above 10°C in winter, resulting in mould, and this drastically affects the bees' health. This is the reason I have for the first time stopped using single-walled box hives.

If your hive has mould on the walls or comb then it's not suitable for bees. If you look at the comb inside a natural tree cavity after winter you will see it is in pristine condition, gilded in places with a red trim of propolis. The tree provides not only 5x the insulation of a singled walled hive, it also provides thermal regulation because of the mass of the tree. The biggest tree hive I have seen is in a forest in Nottingham, here the girth of the tree is 10m and has a weight of 23 tons!

Two points to note here are that the rougher the wood, the better it is for the bees: more propolis opportunity. Also, I have used end grain on the plugs facing into the hive to allow humidity regulation of the bees. You can blow through the wood plugs as wood is made up of thousands of small capillaries. In a tree the top and bottom of the nest would be end grain.

Because pallet wood comes in all different shapes and sizes, it's not possible to give a cutting list. You must use what you have and make that fit with parameters bees seek in the wild:

- Volume ~ 40 Litres.
- Internal diameter ~ 24cm.
- Walls at least 5cm thick.

These simple parameters are based on:

1. The work of Thomas Seeley on the preferred cavity sizes of Bees.
2. The internal diameter and thickness of Zeidler tree hives, a form of tree beekeeping which has a successful 2000-year history.

If your hive is based on an octagon design, I have found that a 10cm plank of about 90cm in length is perfect for the internals of the hive. But beware, this hive has an inner and outer sleeve, with a gap of about 1-2cm between them. If the inner core uses ~10cm wide planks, then the outer sleeve will need planks at least 20.3cm wide, and that is not common for pallets (only 2 of my 10 pallets were of this size). If you have smaller planks you will need to make polygons with more sides and this is where an online polygon calculator like calculatorsoup.com is invaluable.

It's worth spending some time planning this out using the 'measure twice, cut once' principle. It's amazing how this project transforms humble pallets into precious resources.

Making The Core

Once you have the planks you can make the core:

Figure 4 Creating the core

Here I glue planks to the floor and top blocks, temporarily holding them in place with screws while the wood glue dries.

Some notes about the core:

- Did you remember the plugs take volume away from the hive when calculating the core plank lengths?
- The planks can be straight-edged, this saves time.
- The planks can be really rough - the bees will thank you.
- It does not matter if there are a few gaps.
- It's a good idea to recede the plug very slightly (~0.5 cm) into the core, making a 'microcavity' between the plug and an eventual outer roof or floor. This aids in the humidity regulation mentioned earlier. (Figure 4 image does not show this, but I'll mention the microcavity again later.)

Once the core is made it can be wrapped in hessian (burlap):

Figure 5 The completed core wrapped in hessian

The hessian is stapled on and holds the core together. It also separates the bees from the sawdust insulation which will be added in a later stage.

The Outer Shell

The next step is to make the outer shell. Aim to have 1-2 cm gap to the core for insulation. Use the polygon calculator to get the dimensions you need for the core size you have, and if your planks are not wide enough add more sides to make a bigger outer shell.

Figure 6 Outer shell being glued

Here is the outer shell being glued up with waterproof wood glue. For the outer shell, you need to bevel the plank edges with a table saw. Set the table saw angle to 22.5 degrees for an octagon. I use the same straps for securing the hive to the tree to also hold it together while gluing.

Figure 7 Corner staples

Gluing eight planks of wood together is hard... but if you staple them together first before applying the straps it can make a difficult job very easy.

Figure 8 Core inserted into shell

Add the core to the shell and fix them together with a ring of long screws around the top and bottom. This fixes the shell to the core and maintains the 1-2 cm insulation gap. At this point I recommend you read page 14 for details on preparing the core and shell for the entrance holes.

The picture above has jumped ahead a bit and shows the bottom detail.

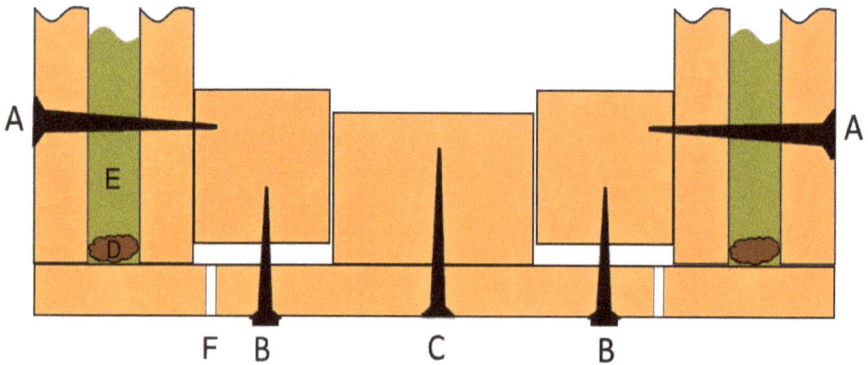

Figure 9 Bottom Detail

The key points about the bottom are:

- The distance between the outer core and inner core is set by screws (A) on each face of the nest.
- Use a large hole saw to cut a human entrance into the hive's bottom plug. This is for cleaning and inspection purposes Screw C fixes the plug to the floor plate.
- I use some spare oak planking to make a 2cm thick top and bottom. While this can block off the end grains of the plugs, it's easy to create small ventilation/drain holes (F) to the end plug 'microcavities'.
- The inner and outer cores are filled with the table saw sawdust (E).
- Use hessian (D) to seal the bottom of the insulation gap.
- The floor and attached plug can be removed by unscrewing two screws with bolt heads (B). The bolt heads just make regular removal easier.

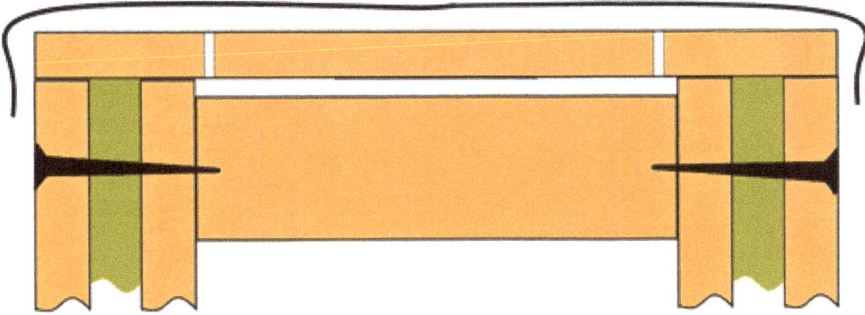

Figure 10 Top Detail

The top uses the same principles as the bottom but has a much simpler design. I have kept the top plug sealed/immovable as this simplifies water-proofing. While you can attach old brood comb to a removable roof plug to attract a swarm, the traditional reason for this was to guide the bee comb building so it is parallel to the tree hive's human side entrance, making honey harvesting easier. However, to attract bees it is sufficient to put a couple of drops of organic lemongrass oil onto the hive entrance and smear the inner walls with old honeyed comb from the bottom. I use an easy spreading mixture of beeswax melted in warm olive oil to cover the insides and entrance holes to the nest. Once cooled this makes a smooth paste that is easy to apply.

Figure 11 Finished hive waiting for a tree!

Two 30mm entrance holes have been added, 1/3 and 2/3rds down the hive. While I understand lower holes equal less heat loss, bees must also get moist fresh air close to the brood. They like brood nests near entrances for a reason. As the brood field gradually moves down during the year, the two holes facilitate this. I find that bees like to use the two holes for different purposes as the year progresses. I often see them use the top entrance for expelling heat and moisture and the bottom for bringing in dry fresh air. The holes are small enough for the bees to control the size easily with a propolis curtain just as they would with a knot hole in a tree.

Behind the holes is a wood plate that replaces the insulation gap. This allows a clean hole from the outer shell to the inner core. The roof is covered with a simple roofing rubber membrane, stapled on. Rubber is preferred to metal which can be too noisy in the rain, and long-lasting wood roofs are difficult to make and keep waterproof.

The hive is painted with a non-toxic 'Scandinavian' paint based on a metal oxide. Unpainted pallet wood would not last very long otherwise.

The hive can be strapped to a tree at a bee-friendly height of 2-4m. Ideally, the weight is supported by a natural branch formation, or a wood step could be bolted/rebarred into the dead heartwood of the tree with little impact on the tree. Ratchet tie-down straps can hold it in place, but these should not strangle the tree. Use two or three small blocks of wood to stop the supporting straps from cutting into the tree.

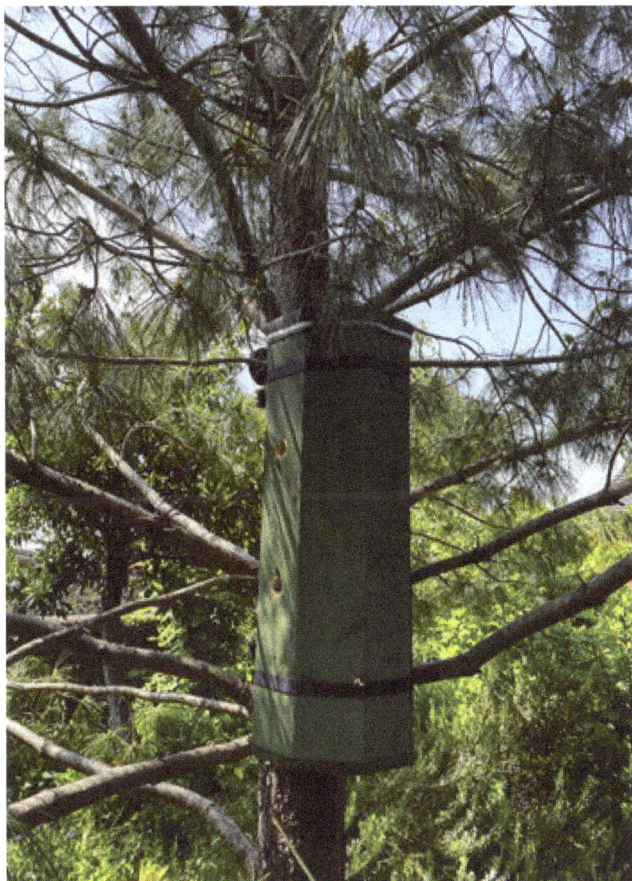

Figure 12 Pallet hive in a tree being checked-out by scout bees

As the hive does not have the mass of a tree it is best to attach it to the shaded side of the tree to avoid overheating, with the entrances pointing to the sunrise position. Bees tend not to care which direction the entrance points, but not directly towards the prevailing weather is a good idea.

Honey bees are the first active pollinators in early spring time for many trees and flowers, and nests should be locate in areas with good sources of early pollen and nectar. They should also be widely spaced with at least 500m between nests, just as in the wild. This spacing greatly reduces the spread of disease between nests and the impact on other pollinators. As the nests will not be harvested for honey and because of the small size and thermal properties of the nest, the bees will consume 10x less nectar than a large commercial hive, again ensuring their impact on other pollinators will be balanced and proportionate.

Making a pallet hive can be tricky, but very worthwhile. It's important to be flexible with the design and work with what you have. I have since thought of small improvements for next one, but I'm happy with the basic design. It's already my favourite hive. Based on the design of this hive I have already seen a screw less version from the UK and another variation from Switzerland.

But Does it Work?

The nest was up for just two days before a swarm of bees found it and moved in! Bees use a voting system to select nest locations, so I take that as a solid vote of confidence.

Figure 13 There is nothing I need to do now, except watch.

Why Make Pallet Hives?

I have come to understand the main reason we do not have many wild bees is due to there being so few natural tree cavities. In ancient forests, like the forests of Boughton House, Blenheim and the New Forest, dead trees are left in place, and there are many cavities and many free unmanaged honey bee colonies. Here the bees find homes that are an exact match to their preferences: small warm cavities with small entrances deep inside large heavy structures. Ancient forests are rare - stupidly we are chopping them down in the UK to make way for high-speed train lines to save a few minutes commute time. Nature cannot wait 500 years for replacement forests to grow.

While nothing can replace an ancient forest habitat, installing pallet hives does give back to bees some of the opportunities they need to thrive. We are giving the bees the maximum agency to be who they are. This hive is not for honey harvest - it's for the bees and our joint future together.

Bees thrive when they are wild and free, and they remind me in a time of lockdown how precious our freedom is. Let's make sure after lockdown we value our freedom even more.

www.ingramcontent.com/pod-product-compliance
Lightning Source LLC
Chambersburg PA
CBHW040155200326
41521CB00019B/2609